The Universe in Black and White

....

Printed and bound in China by Imago

Hierophant Publishing
8301 Broadway, Suite 219
San Antonio, TX 78209
www.hierophantpublishing.com

If you are unable to order this book from your local
bookseller, you may order directly from the publisher.
Library of Congress Control Number: 2012935827

ISBN 978-1-938289-01-9

The Universe in Black and White

A Plain and Simple Illustrated Guide to Time, Space and the Meaning of Life

Terry Favour

Hierophant publishing

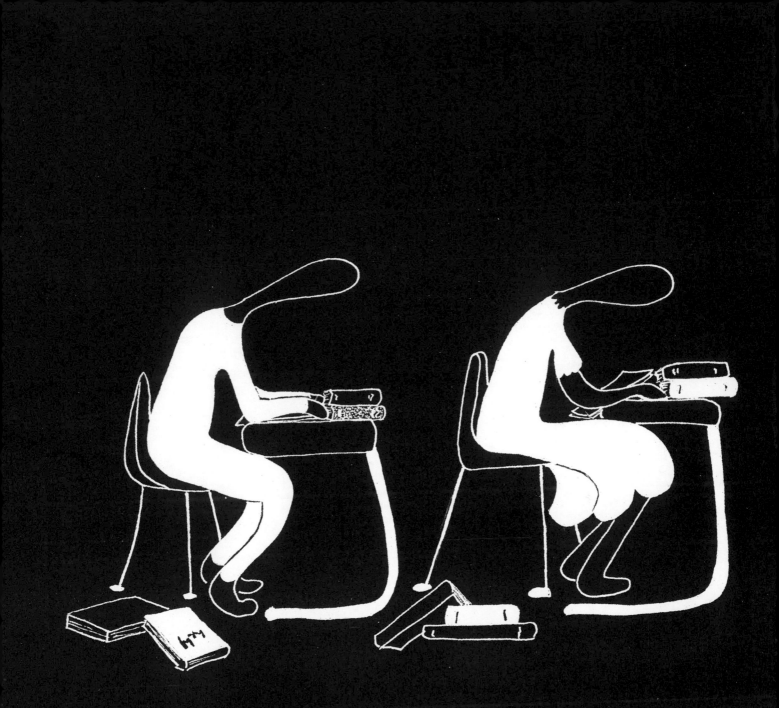

Dedicated to 21st-century alchemy
and all those souls joyfully participating

Introduction

This book first came to me in the form of a vision. Its simplicity struck me as profound. But when I tried to capture my vision in words, it proved to be more elusive than a kite without a tail. So I gave it up.

My children are grown now, but ever since they were quite small, we have had many wonderful conversations about our changing world. It was obvious to me that the environment they were growing up in was straddling more than one reality. At times they were being asked to make very contradictory decisions. I knew about contradictory worldviews. I was born into a Christian Science family, and when I started school, it took me no time at all to discover that I would be looked upon as a freak if I voiced any ideas from home.

Since I could not reconcile these very different worldviews, I had no choice but to discover the truth for myself. That reconciliation, as many of you know, is a lifelong endeavor. So, as my children grew up, we pursued these discoveries together, and we do so today.

Now, my children and I often talk about quantum physics, because it may have impacted our present worldview the most. Recently they got very specific with me, asking me to lay out the very simplest explanation I could. They said they could grasp the ideas we spoke about, but it was difficult to repeat them to anyone else.

Suddenly there it was! My vision poured itself onto paper as easily as cream pie. My family and friends helped me poke and prod the words, just a bit, into this book.

That is how this book came to be, bit by bit. However, bit by bit may not be the best way to read it. Simple or not, these are difficult and relatively new ideas. A key to understanding this particular approach is to read the book slowly and carefully, and — most importantly — if possible read it straight through, all at once, cover-to-cover.

I wrote more as my own understanding of this unfathomable subject deepened, I became astonished at the amount of information available to us from our own Western traditions. Like many others, I had looked to Eastern spiritual traditions for the much-needed guidance that seemed so lacking here in the West.

The truth that is the result of the past 3,000 years of concentrated effort has just in the last 100 years come to fruition. Because this fruition has just come about, and, because most of the people responsible for it do not themselves understand what has transpired, this magnificent culmination has gone unnoticed.

In fact, in the 3,000 years it took to accomplish this tremendous feat, we actually forgot what it was we were searching for in the first place. No wonder we don't recognize the result.

We forgot that when we began this quest for knowledge, we were asking, "Who are we?" But as we approached the scientific revolution, the quest became about our own power. Francis Bacon, who was so influential in bringing about the scientific revolution, made it clear that the only knowledge that is not "mere trifling" is the knowledge that enables us to make nature do our bidding. In spite of that, using the fantastic technology that Bacon aspired to, we have stayed on course, as our Greek forefathers would have wished. And we have answered the question, "Who are we?"

As yet, most of us don't realize this. But one man, living almost the entire span of the twentieth century, was very clear about this all-encompassing Western adventure. His name was Owen Barfield. We sometimes hear about him in conjunction with his friends, C.S. Lewis and J.R.R. Tolkien. Owen Barfield was a linguist who made his discoveries about our Western quest from that discipline. He wrote the book *Saving the Appearances*, which, though it is beautifully written, is difficult to read. Despite the difficult reading, his succinct and brilliant overview of our Western evolution of consciousness makes his book one of the most important to date. While I have not written strictly from his point of view, a portion of this book was primarily written in an effort to make his great work more accessible.

In no way should any part of this book take the place of *Saving the Appearances*. It should be regarded as a kind of groundbreaker for those who are interested in pursuing further what I believe to be the end of an era and the beginning of a new one. This era could be the beginning of the most unimaginable and exciting adventure ever dreamed of.

With every day that passes, the self-imposed peril that threatens our planet and our humanity becomes more apparent. We have come a long way over the ages. We have at our fingertips all the information that we need to evolve, once more, into the potential that awaits us.

One way or another, undoubtedly, the creativity that makes up our cosmos will take the next step. What I don't know is whether our human forms will take this step willingly, or even successfully. Like spoiled children, we may only step forward after causing ourselves and our planet much pain.

I am just an ordinary person. I am not a scholar, not a spiritual teacher, not a scientist. There is available right now a lot of information provided by qualified people. Armed only with an open mind, the average person can educate himself. We all can become aware of the magnificent discoveries that point the way through this dangerous labyrinth.

It seems that this job has been left up to each one of us, individually. There is no one person or teaching available to save us. Perhaps the maturity necessary to find our own way is embedded in the transition itself. In any case, it has been my good fortune to come across the heartfelt written words of caring scholars, spiritual teachers and scientists. These people are humble folks, reaching out to us, trying to impart the wisdom that their in-depth studies revealed to them. Over the years, their thoughts have become the fabric out of which my own understanding is woven. It is a weave in which I can no longer find the individual threads. I cannot list all these wonderful sources.

This silent tapestry, made from all that wealth of wisdom, whispered to me as I moved around my daily life. Eventually it expressed itself in the little chapters that make up this book. It is my hope that, in some small way, they can contribute to the monumental effort it will take to free ourselves from the bondage of this worn out paradigm, and propel us safely towards our destiny.

Terry Favour

What is a paradigm?

A paradigm is the totality of how a person or a culture views reality.

In this book, I am describing a view that is generally held by most people in the Western (as opposed to the Eastern) part of the world.

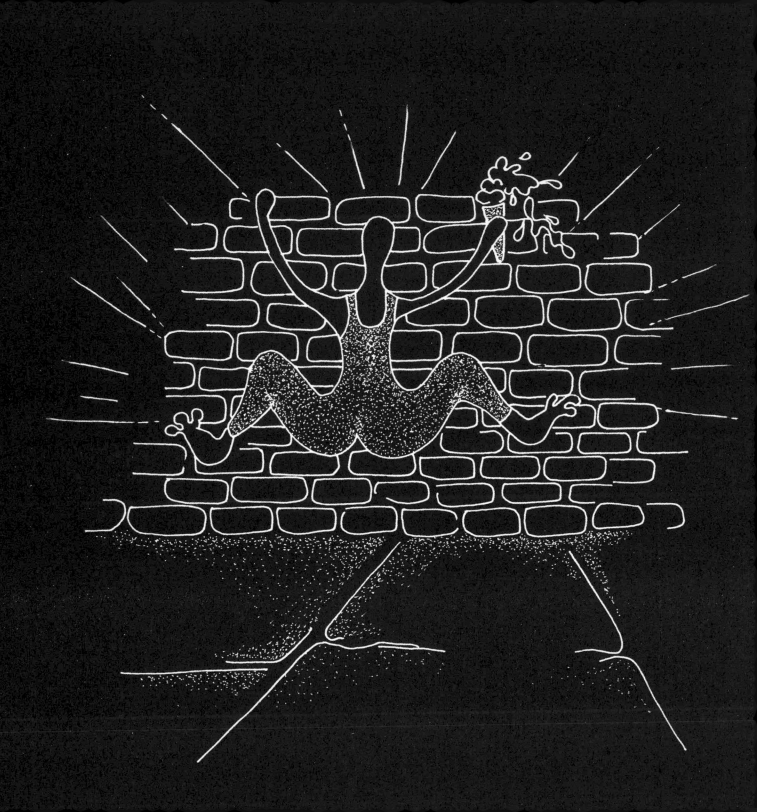

This view illustrates the limits of our physical and psychological worlds that can be documented by science and understood by the average person.

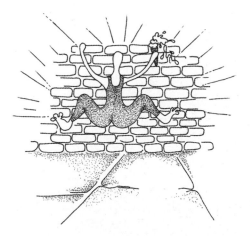

Our current viewpoint began more than 2,000 years ago, with ever increasing and more complex probing into and documentation of the natural world.

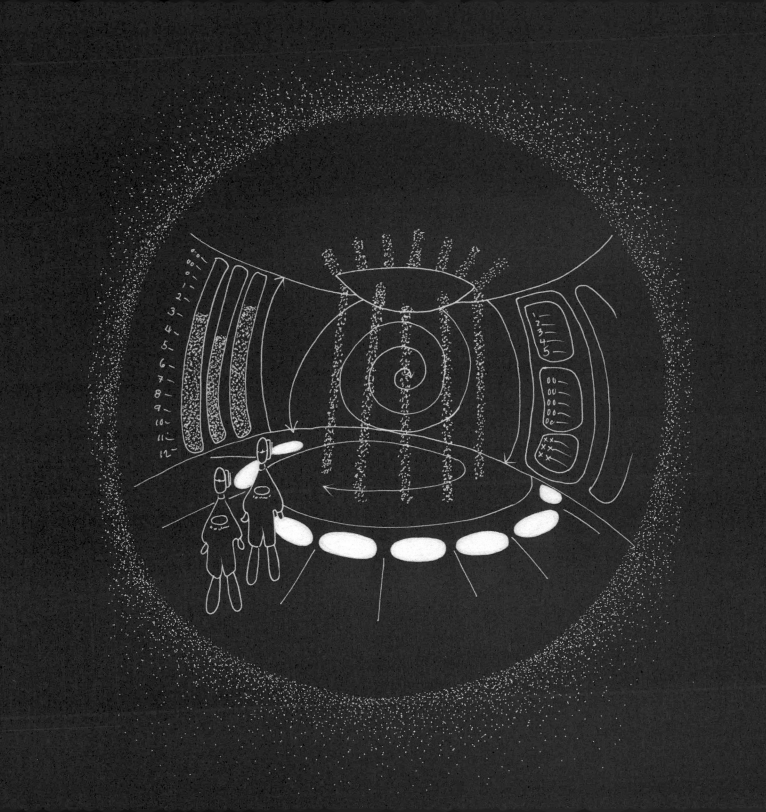

Today our technology is extremely sophisticated and has made some discoveries that are so startling that they must be termed replete, and therefore defined as **paradigmal**.

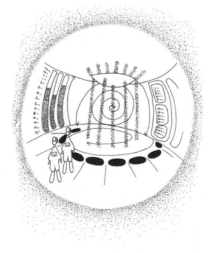

How a Paradigm Shifts

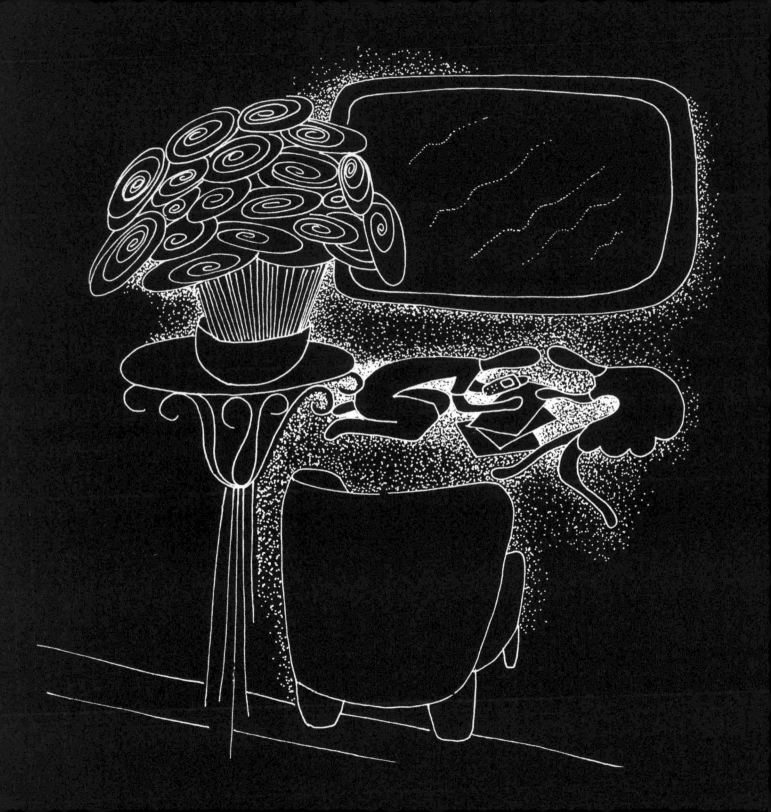

To further describe how a change in paradigm happens, imagine that you have just walked into a friend's living room. There is a bouquet of flowers on a table but you don't notice it. You are more interested in a big-screen TV that your friend has just purchased. You and your friend spend several hours figuring out how to program this new technological wonder to suit your friend's needs. Never once do you notice the bouquet.

If you leave your friend's house never having noticed the flowers, in your world those flowers don't exist — because your intent and attention were focused elsewhere.

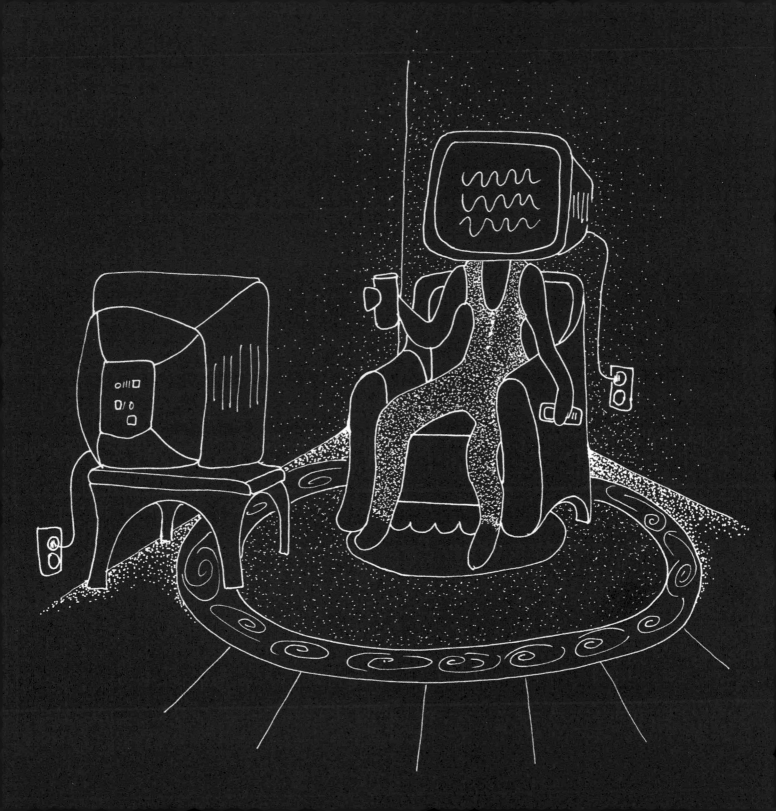

When a paradigm shift occurs, it is
not because that truth is new, but because
it is a truth that has gone unnoticed. One's
intent and attention have been focused elsewhere.

For more than 2,000 years, the Western mind has been focused on the big-screen TV and how it works. In other words, the Western mind has been examining the nature of the substance we call matter, its properties and its laws.

Has this focus
missed something?

Keeping in mind what has been presented, let's examine dualism — because its study will point toward the answer to this question.

What is duality
and what has it to do
with you and me?

It is very important to our discovery of what has been missed, that the nature of duality is understood.

Dualism involves the old idea that
we can't know joy without sorrow, that
we can't know hot without cold.

Imagine that you live on a planet that has 98.6-degree weather everywhere all the time. In fact, nothing on the planet has a temperature other than 98.6. No one there has ever heard of or experienced any other temperature.

Neither you nor anyone else would have a concept of cold. For that matter, you wouldn't have a concept of hot either. In fact, on your planet there would not be a concept of temperature, or of 98.6 degrees, at all. The whole issue of temperature would be unknown to you.

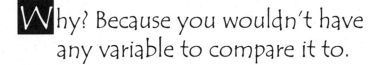

Why? Because you wouldn't have any variable to compare it to.

No concept can exist without comparison. Things must have contrasting characteristics or they are not distinguishable. If only joy exists, then the components making up the concepts of joy and sorrow do not exist.

The Heart of
the Question and
the New Paradigm

This need for contrast brings us
to the very heart of the question,
and to the new paradigm.

Science has come to a major crossroads, after more than 2,000 years of examining things.

First, humans examined objects
that anyone could see or feel, things we could
see with our eyes or touch with our hands.

Then scientists began to build instruments so that they could penetrate the surface of those things we could see or feel.

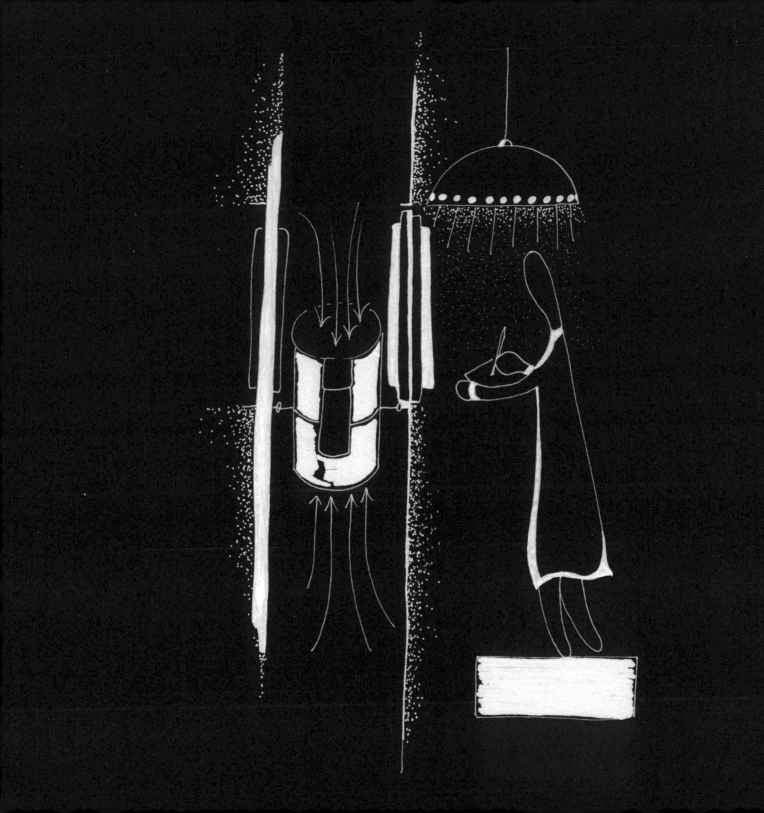

Later, with more sophisticated instruments, scientists began to examine the very material these objects were made of, thinking that they would eventually come upon their basic components.

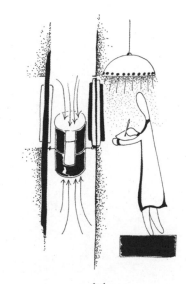

Scientists called this material matter, and for a while thought its basic component was the atom.

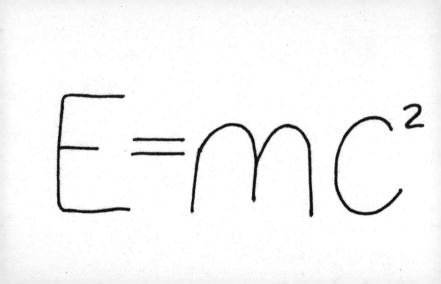

But then science split the atom
and found energy.

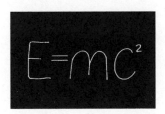

Upon close examination, scientists found that they could not find any, in the ordinary sense, stable or predictable characteristics of that energy.

You may have heard that energy sometimes appears as a wave, and sometimes as a particle, and that the science addressing this is called quantum physics.

Because quantum physics involves
a lot more than is addressed here, understand
that I am simplifying very complicated matters.
But we can say that the search for the smallest
components of matter must STOP here. If matter
turns out to not be made of 'any-thing' with
measurable qualities, thus making it a
'no-thing', the search for the SOURCE of
matter must have its beginning here.

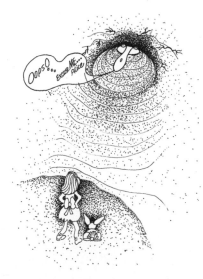

So what then is the source of matter?
That is the most important aspect of our inquiry.

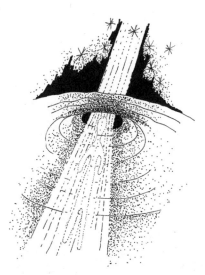

B ut wait! Were we not considering
how no concept or thing could exist without
comparison or dualism? Could this hint at where
to look for the source of matter?

In our more-than-2,000-year
search for the truth of our material source, did
we overlook the fact that to examine any thing
at all, we depend on the very real reality
of no-thing as well?

Here again, we must look at our temperature analogy. We can see that temperature is an entity made up of cold and hot. When cold and hot are indivisible, as in the case of our planet with only a 98.6-degree reading, humans cannot notice the concepts hot or cold.

F or humans, dualism is true across
the board. There can be no awareness of a
concept or a thing without something
to compare it with.

So what does this mean, really?
This means that we humans have to examine
ourselves much more closely. We know that
we are made of matter, like everything else.
But since at its core, matter has no substance
that can qualify it as a thing at all, then what
are we? If we exist at this very moment with
no substance as our core (meaning that we are
actually not made of anything), then what are we?

What then is this dualism that is most basic to you and me?

To understand our basic dualism
we have to look again at what science has uncovered.

The source of the material world, the basic building blocks of matter, cannot be found in matter itself. The source of matter is something other than matter, something that is not a thing, something without properties but nevertheless real.
No-Thing/Thing

How are scientists supposed
to consider this no-thing that has appeared
at the heart of their observations?

It seems that science has become
its own worst enemy. The basic building block
of what we call real is made up of something we
have been calling unreal. Since this non-material
is the basic building block of material reality, then it
too must be real. Here, at last, we have come to
the first component of the paradigm shift …
the reality of no-thing.

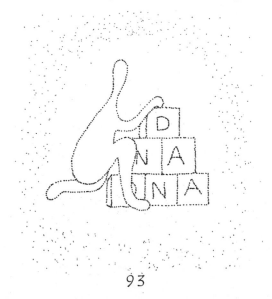

Remember, a change in paradigm means finally seeing something that was always there, but was not noticed.

Because of this shocking new shift
in scientific discovery, we are being forced to
notice that there is much more to reality than
we ever imagined. This shift reveals that reality
includes something we cannot imagine,
or touch, or even think about.

This new view involving the reality of **no-thing** describes one element of the paradigm shift.

At this point, we must turn the light of inquiry on ourselves. As mentioned before, if science has shown that the basic building blocks of our material reality are made of the non-material (something with no properties), then you and I are not material either.

What we come to when we probe into
the matter that makes up our bodies, is something
non-material, something without properties.
Yet, something, most certainly, **does exist**.

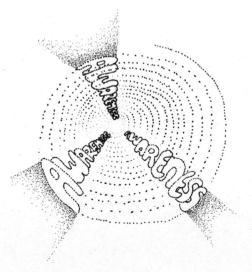

This is the dilemma — the shift
of paradigm that allows us to see what we have
not been able to see before. We must see, in an
indirect way, the evidence of that which exists
but has no properties that we can measure in
the usual scientific way. Remember that dualism
is critical to our ability to think and form
conceptual ideas, to be conscious.

How is it that we can indirectly see
that part of the duality that constitutes the no-thing?
We can see it by using duality itself, but most
importantly, by noticing that we are this duality.

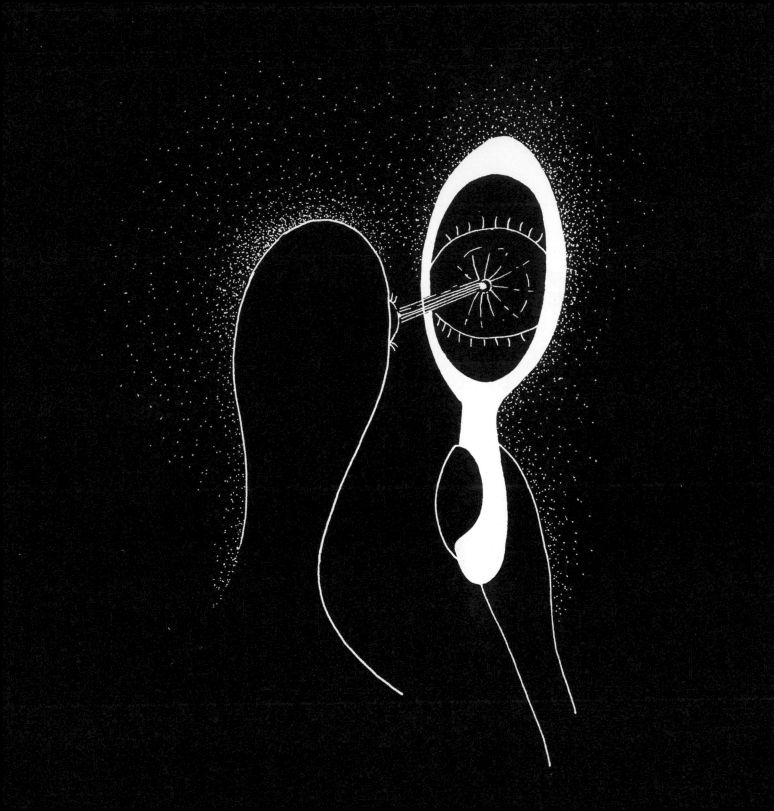

We are this dualism. I cannot see
my own eye with my own eye without the use of a
mirror. But, I am aware of my eye seeing. The dilemma is
that we are not aware of being both a body existing in
time and space, and at the same time being the source of
that body, and as the source, beyond time and space. The
source, like the eye, can see its own body and all bodies,
but also like the eye, cannot see its own true self.
It can only be aware that it sees.

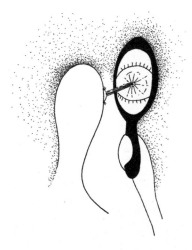

The Conclusion
of the Paradigm Shift

Until now, scientists have studied objects they considered to be located outside of themselves. Remember learning about the scientific method — based on objective rather than subjective study? With this new mind-altering information, ironically, information arrived at using the scientific method, scientists have discovered this dualistic dilemma. Human bodies are made up of matter, but the source, the building block of matter, is something that is non-material, no-thing.

Even though we find the source of our own bodies' makeup to be non-material, we must recognize that this non-material aspect that we now see that we are, knows that it exists. This is BIG! This means that this non-material is conscious — not empty nothingness, but rather, full, overflowing, intelligent no-thingness. This means that the deepest core of each of us is an awareness, and that awareness is our true self. Like the eye, it knows that it exists, and, like the eye, it cannot directly see itself because it is the source of all that it sees.

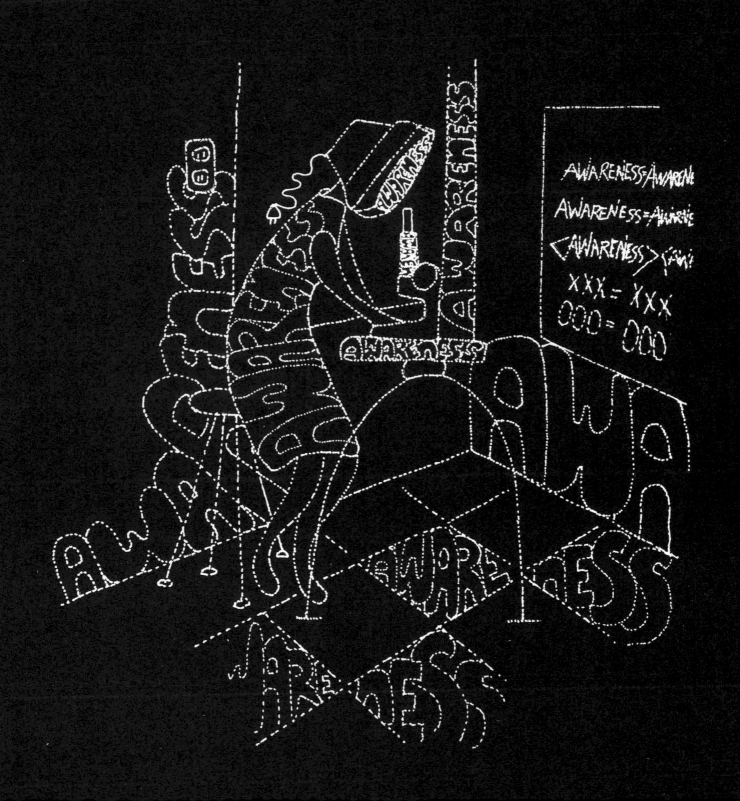

Let me be very clear. It must be recognized that what is being studied (the so-called object), and what is studying the object (the scientist himself — or the so-called subject) are at the core, the same thing. They are two parts of one whole. To put this in a way compatible with what has been presented thus far, we would say that the object being studied and the scientist studying the object are the dualism required for consciousness to exist. But, in actuality, that is a very clever trick, an illusion, just as Eastern philosophy says it is. This trick, this illusion of dualism, allows the self to be aware of itself. But it is not scientifically true. The object and the subject are not two separate things.

Look around you. Look at the absolute precision of nature. Look at your body, at other bodies, at plants, trees, the oceans, the sun, the moon, the entire cosmos. What we see is integrated wholeness, invisible precision that is intelligence itself. That same silent unseen intelligent awareness is the core of you and me, our true selves. Our material bodies are the products of our true selves, not the source of them.

For the Western mind, it is more than unsettling to have the rug of objectivism pulled out from under it. In fact, it is paradigmal. Like it or not, it has happened. The Eastern spiritually oriented mind has always been in accord with this non-dualistic view. Now this view is documented scientifically in the West, bringing two great inquiries into the nature of our existence together for the first time. Has this melding of two worldviews happened? Not really.

What will it look like when it does happen?
I will explore that question next.

Paradigm Review

We have now acknowledged
that our solid-looking world is not solid.
We followed the cutting edge of Western
science to quantum physics and the discovery
that things, at their core, have no real
material substance.

We discovered that if this quantum
view is the true foundation of our world, then
our human bodies, at the core, are not made
of any real substance either.

If we are not material substance, what are we? We came to the conclusion that we are actually AWARENESS. We are the awareness that the forms of the material world appear to be solid WITHIN.

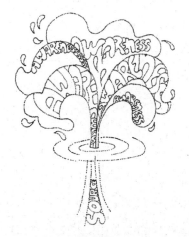

Let's try to uncover what not being material substance really means, and therefore, how the experiences of our daily lives actually happen.

The focus of our attention creates our reality by including some things and at the same time excluding other things. Our focus of attention continually creates limits. The very definition of form includes the idea of limits. Without limits we are faced with undifferentiated potential, with waves and particles, and with the realm of infinite possibilities.

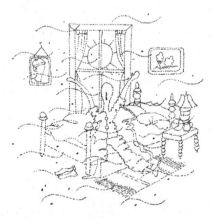

If this is true, how do things get to be things? In other words, how do waves/particles become form? How do they get to be familiar objects?

Do things exist as solid matter if they go unnoticed? Do things exist independent of US? Is there some interrelatedness between what appears to be solid matter and OUR senses and OUR minds?

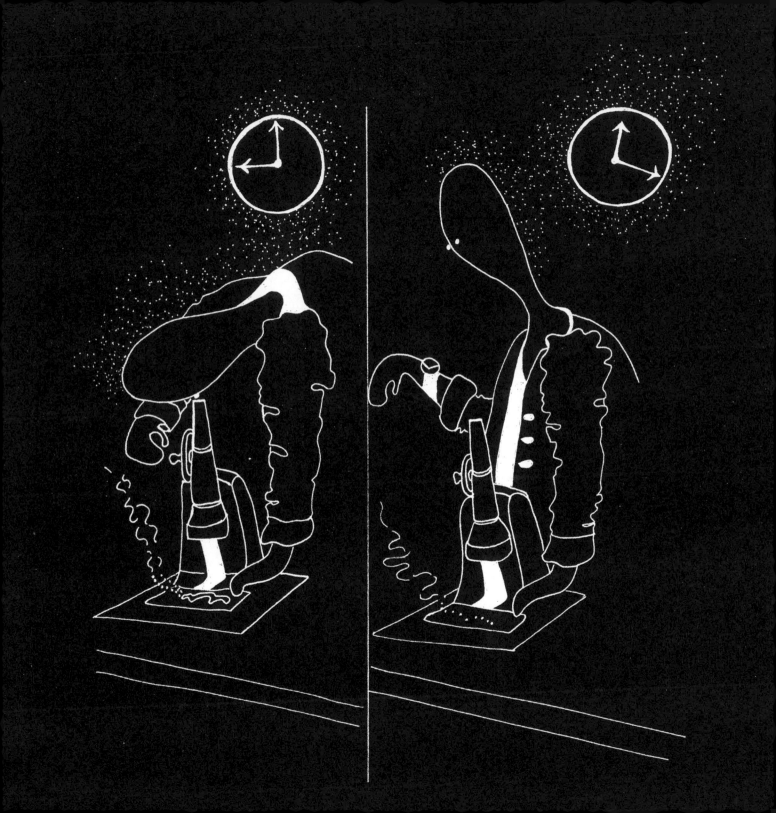

In very simplified terms, quantum physics (which studies the very small), says that a wave equals potential until it is observed. Only when a wave is observed does it go from potential to actual, from a wave to a particle.

Focus of Attention

To make this clear, we need an analogy. Consider the rainbow. Does a rainbow exist independent of its observer?

Think of all you know about a rainbow.
For a rainbow to exist, three ingredients must
be present. There must be sun, raindrops and, of course,
the vision of the observer. In other words, a rainbow's
existence depends on the actual optical apparatus of
the eye. It is that apparatus that puts the sun and the
raindrops together, forming the 'vision' of the rainbow.

E ven so, I still can't touch the rainbow.
If I try to walk to the end of it, it won't be there.
But if a friend observes it with me, she will see it
too, and we will agree that it does exist.

Then we must consider this. Is it true
that as soon as anyone sees a rainbow it therefore
exists? If so, what is a hallucination or the vision of
a madman? Would you believe a person who claimed
to have seen a rainbow on a sunless or cloudless day?

For us to know that a rainbow is actually there, there must be sun, raindrops and the observer's vision. And somehow the observer's mind is also involved. There must be collective agreement about the existence of rainbows, as well. But let's not go too fast. We will investigate this further as we go along.

Take the rainbow example and apply it to a table. Like the rainbow, we can see a table. But unlike the rainbow, we can also touch a table. It makes a sound if we hit it. If it is made of wood, it could burn and then we would smell it too.

Unlike the rainbow, our senses tell us
that a table is made of solid matter. Just as you
did with the rainbow, think of all you know about
a table—which appears to be made of solid matter.

We have considered the effect an observer has upon waves, and that in the realm of the very tiny, waves actually become particles when observed. Matter is composed of particles. The table appears to be made of solid matter.

But NOW we know that even a table is
not solid. What actually makes up the table is what
I have been calling particles: waves that have become
particles upon observation, particles that have interacted
with sense perception and then become the table. Neither
the raindrops nor the particles of the table are like what
they become. Just as the rainbow is the outcome of the
raindrops and my vision, a table is the outcome of the
particles, my vision and my other senses. And now it
is time to bring the mind's role into this.

To explain how the mind fits in, imagine a situation. One morning as you awaken, the room you are in and its furnishings are just coming into focus. You notice it all, including a table. You don't know exactly where you are. You recognize objects in the room but they seem totally unfamiliar. You realize that you are NOT in your bedroom at home! Suddenly you remember that you have been traveling. You remember where you are. In an instant everything snaps into place. The room suddenly has context and looks familiar.

What happens though, when something appears to the senses for which the mind has no context—collective or personal? For example, have you heard the story that when the first European ships sailed into the harbor of some primitive island, the natives could not see the ships at all? The ships were outside their experience, and therefore outside the perception of their senses.

How does this tie into the beginning
of this book and the paradigm change it disclosed?

When humanity agrees collectively about what is real and what is not real about the world, then we have a paradigm. When, in time, that view changes, and the human agreement changes, as it has many many times, then the world changes too.

Just how far does this collective agreement go? Would there be form at all if there were not a sensate being with a conscious mind? This book proposes that there cannot be form as we know it without human observation, since the manifested and the unmanifested are different aspects of the same thing. And what of collective agreement? If we all agree on what is real and what is not, have we not shown that our concept of reality IS our agreement? Do we not call this agreement FACT? Once it was FACT that the world was flat.

If the collective agreement changes, then
the form actually changes. This can be quite radical.
But once the change has occurred, the past paradigm passes
out of experience to such a degree that the true meaning
of the prior view is lost. It becomes inaccessible, since
it is not WITHIN the particular confines of
our current view.

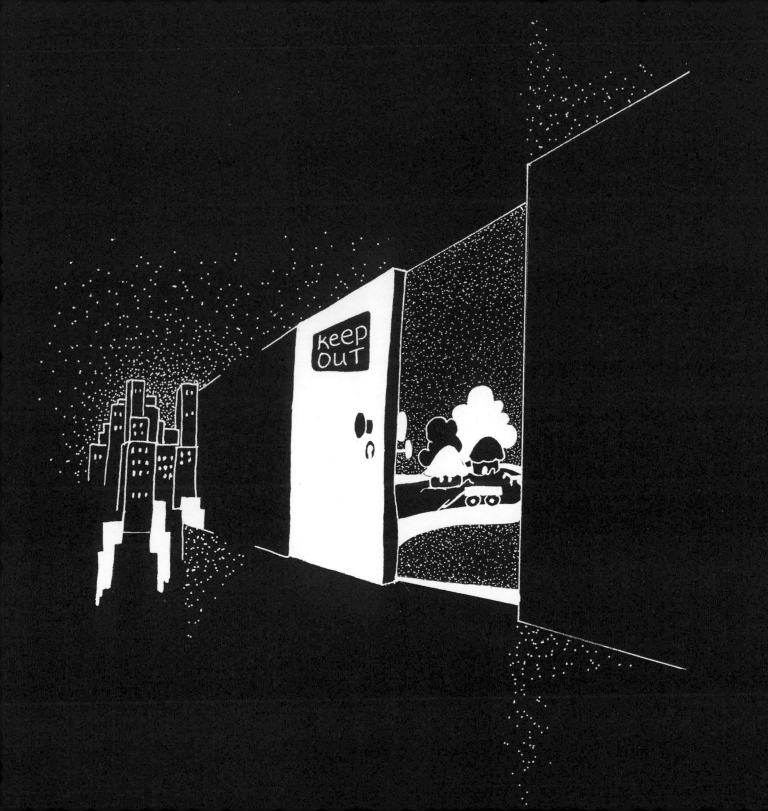

This is our predicament today. Something happened about 500 to 600 years ago that radically changed our paradigm, but we have no memory of how we actually viewed the world before that change. The limitations of our current collective agreement lock us out.

Two things happened over time about 500 to 600 years ago. One was the invention of the printing press, which spread the written word. The other was the scientific revolution. In order to better understand these changes, I introduce the use of three words in a particular way.

Three Words

The three words, originally proposed by linguist Owen Barfield in *Saving the Appearances*, are: alpha-thinking, beta-thinking, and most important, figuration.

The most relevant to this discussion, figuration, describes how things become things.

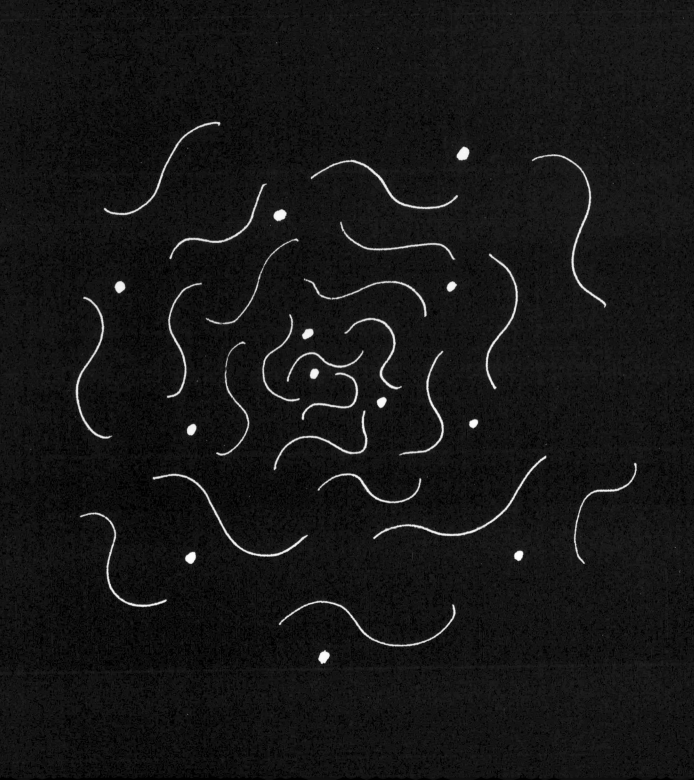

If we continue to develop the line
of reasoning begun on previous pages, we would
have to say that ONLY the waves are independent of
our sense apparatus, until the waves interact with
our sense apparatus to form particles.

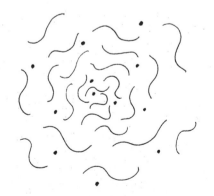

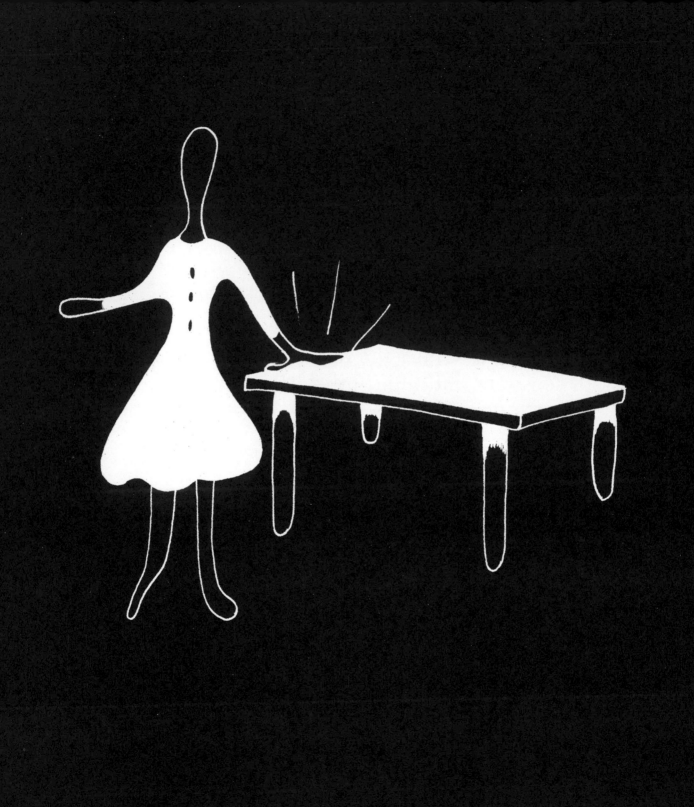

And so, in order for form to arise, "the sense-organs must be related to the particles in such a way as to give rise to sensations …. "*

* *Saving the Appearances: A Study in Idolatry*, Owen Barfield, Wesleyan University Press, 1988, page 24.

185

Also, "… those mere sensations must be combined and constructed by the percipient mind into the recognizable and nameable objects we call 'things'."*

* *Saving the Appearances: A Study in Idolatry*, Owen Barfield, Wesleyan University Press, 1988, page 24.

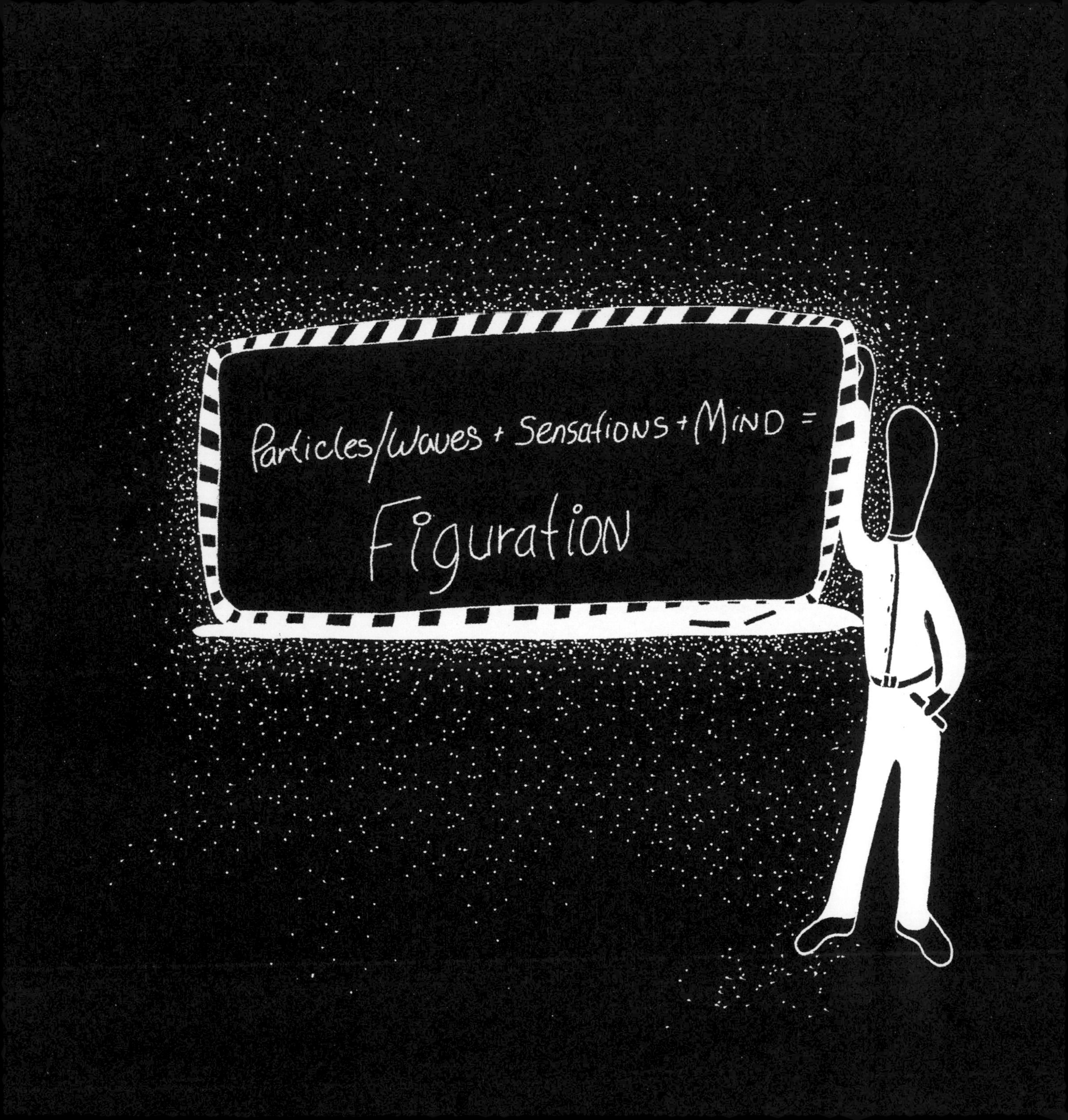
Particles/Waves + Sensations + MIND =
Figuration

It is this work of construction by the mind that Barfield called figuration.

Particles/Waves + Sensations + Mind = Figuration

Barfield used alpha-thinking to describe
how, after we have established objects in our world,
we can begin to think about them.

In the activity of thinking about, as well as in the process called figuration, we remain unconscious of the intimate relationship these things that we are thinking about have to our senses and to our minds.

In alpha-thinking we are even more unconscious than before, for our attitude is to treat things as completely independent from ourselves. We speculate about them or investigate them as originating outside ourselves. We study them in relationship to one another. Scientific thinking would be an example of this.

Beta-thinking is the kind of thinking that we are doing at this moment.

Beta-thinking is the domain of philosophy and psychology. With this kind of thinking we can consider the nature of our collective agreements.

The Collective
Agreement

We are almost ready to look at what happened about 500 years ago. First let us ask what our collective agreement was until that time. What was our relationship to objects long ago, before the printing press and the scientific revolution?

For the most part, wisdom and knowledge were passed down orally, through stories. The depth of abstract thought, the ability to objectify to the extent we do now, was not possible without the written word. People who read and write relate written words to the objects they describe, and also relate written words to each other. They forget that the object they read about has real substance. They are bound up in thoughts and ideas, not real things. They can mentally take part in experiences without participating. That is drastically different from knowing only the sounds of words.

It would be difficult to do what we are doing at this very moment without the technology of the written word. Our collective agreement now is based on it. You and I could not interact as we do without it (beta-thinking). Nor could we do scientific research without the objective memory it provides (alpha-thinking). Since recorded history began, people have used the written word to think about themselves and their environment. However, that did not become the foundation of the collective agreement until most people could read and write, and therefore had access to this tool for thinking.

This written word, this TOOL for thinking, does something more. It enables us to participate with words and our imagination without participating with objects themselves. And that brings us to what our collective agreement was 500 years ago and back through history as far as we can go.

Our collective agreement in the past
was not abstract as it is today. It was PARTICIPATORY,
meaning that there was little difference between our
subjective and objective worlds. Our subjective and
objective worlds were a united whole.

two Sides of the Same Coin

We saw the ground of our own being, and the soul of everything—forms, objects, all of nature—as one united whole. Everything was part of everything else, all acting on and influencing one another. We could only view the world in that way because the ability—this THINKING TOOL—that has enabled us to separate and abstract ourselves from our ground of being, was not in place yet. The written word had not yet been COLLECTIVELY experienced. It was not yet our Western basis for the measurement of truth.

Remember reading earlier that when an old paradigm passes away, it is almost impossible to access it? At this point, I think we need to try. It won't be easy and it needs contemplation. But using our imagination to its fullest, let's see if we can put ourselves for a moment into the shoes of an ordinary medieval man, and think the habitual thoughts common to his paradigm.

As a medieval man, you would not spend much time thinking analytically. Imagine standing on an outcropping overlooking the sea, simply aware that you are looking at one of the four elements of which all things on earth, including your body, are composed.

As that ordinary medieval man, experience
the element of water inwardly as one of the four humors
that make up your very temperament. Experience earth,
fire, water and air as living parts of yourself. You consider
the proper mixture of those humors essential to good
health as well as to good character.

In those times, you would have actually experienced earth, fire, water and air as living parts of yourself. You knew that those elements were also related to the stars, as were you, through the constellated signs of the Zodiac. You saw all of nature as alive, kindled by the same soul force that enlivened you. You saw yourself as equal with, not superior to, all of nature.

SCORPIO

If you had struck a bargain, cut a finger or lost a farm tool, you would have viewed the event, no matter how minor, as intertwined with the whole cosmos. Whatever happened was a sign of your own condition within the whole continuum—that included heaven and earth and all of creation.

Although this example is only a small taste of how medieval man experienced his life, at least it gives us a glimpse of how different the collective agreement was then from ours today. Unlike what we do today, medieval man, and all mankind before him, fully experienced being part of the environment and participating with it.

Our Present Collective

B ut, have we really stopped participating
with our environment? Haven't we seen that FORM
itself depends on PARTICIPATION of the senses with
the environment, beginning at the level of the
very tiny, with waves and particles?

We have also seen how the collective agreement gives shape and context to these forms, even deciding what they will BE by way of focus and attention.

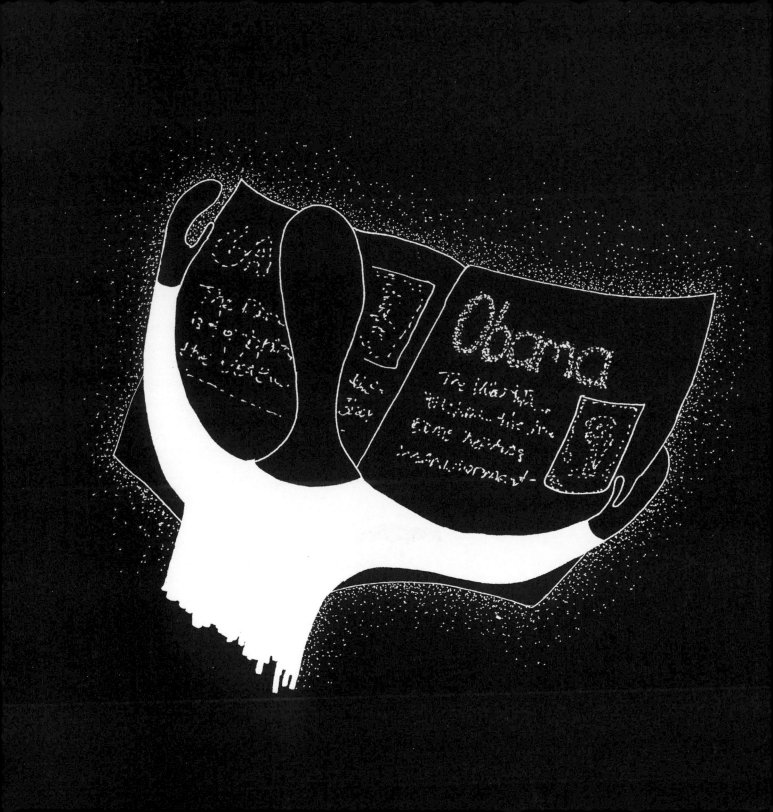

Today the great majority of us know
how to read. Instead of PARTICIPATING
consciously with the four elements as did medieval
man, we have refocused our attention into a profound
PARTICIPATION "with inked marks upon a page."*

*The Spell of the Sensuous, David Abram, Vintage Books, 1997, page 131.

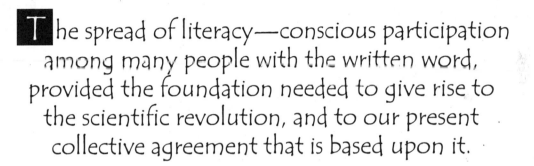

The spread of literacy—conscious participation among many people with the written word, provided the foundation needed to give rise to the scientific revolution, and to our present collective agreement that is based upon it.

Once we could COLLECTIVELY remove ourselves from our environment, that became our paradigm, and all of us began to examine everything objectively. That process of examination itself has become our present collective agreement. In other words, scientific study has become our test of truth.

Comparatively, our collective agreement
is still young. And we are just beginning, with
the advent of quantum physics, to get a glimpse
of what we have REALLY been studying
for the past 500 years.

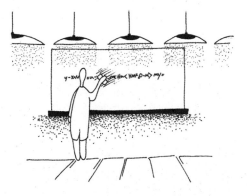

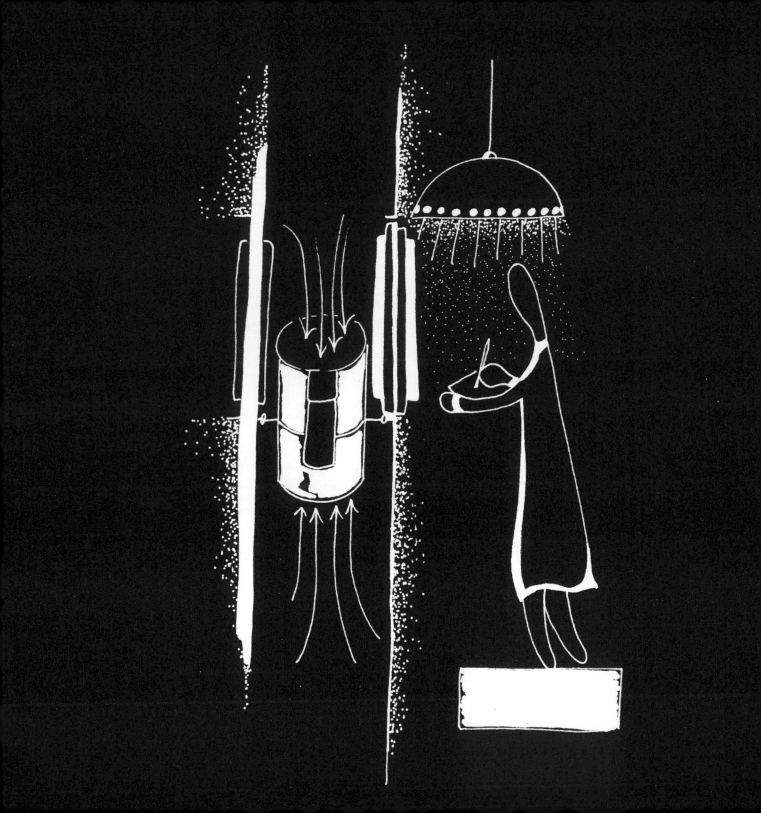

We have not been studying ALL of our human history up until now. What we have been studying is ONLY our PRESENT collective agreement. In other words, what we are beginning to see due to quantum physics is the representational nature of our collective agreements. Our tangible world represents our collective agreements—really just particles interacting with our sense apparatus that are then given context in accordance with those same agreements.

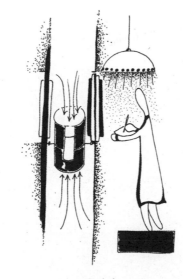

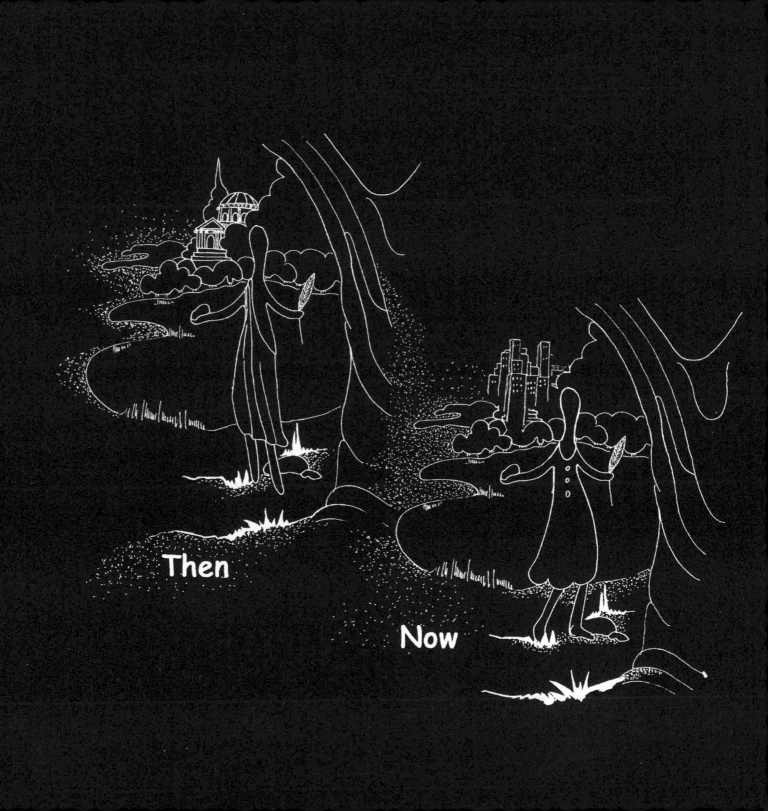

Our present 'representational' collective agreement is based on our scientific outlook. This outlook, combined with our agreement, has us believing that objects and forms are static, and have always been a constant, never-changing reality. Most of us believe that this reality is basically solid and separate and therefore can be studied completely objectively. We believe that objective study, in fact, is how truth is discovered—the scientific method.

Now, TECHNICALLY, science has enabled us to see how we create all collective agreements and the figurations that support them. Science has shown us that we have created our particular current collective agreement and that this agreement is only one possibility among infinite possibilities. Even though our present collective agreement is the fertile ground out of which these discoveries have sprung, this present agreement itself hides from us the real truth of those very same discoveries.

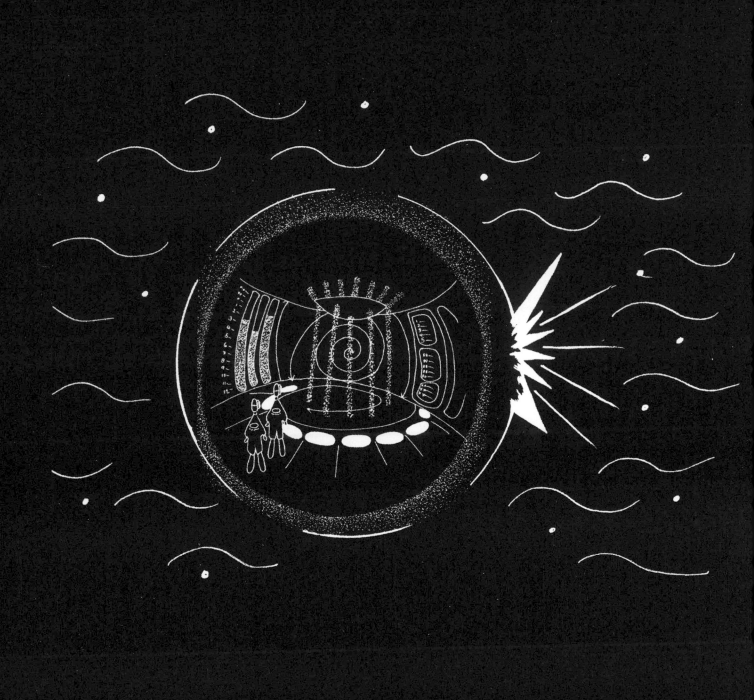

What is the truth? The truth is simply that we have been studying our own creation, our own figuration, and our own collective agreement as though we had nothing to do with its existence. In fact, we have gotten so technically proficient, so scientifically advanced, that we actually broke through our present figuration to the 'stuff' that makes all figuration possible. It is that deeper realm that quantum physics addresses. Yet, for the most part, we don't fully realize what that deeper realm—that 'stuff'—is.

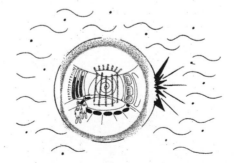

And what is that 'stuff'? It is Infinite Potential.
It is Intelligence. It is Unfragmented Wholeness, and
it is Awareness. It is the foundation of all figurations,
and is that in which all figurations reside. We could
call IT the PRINCIPLE of existence.

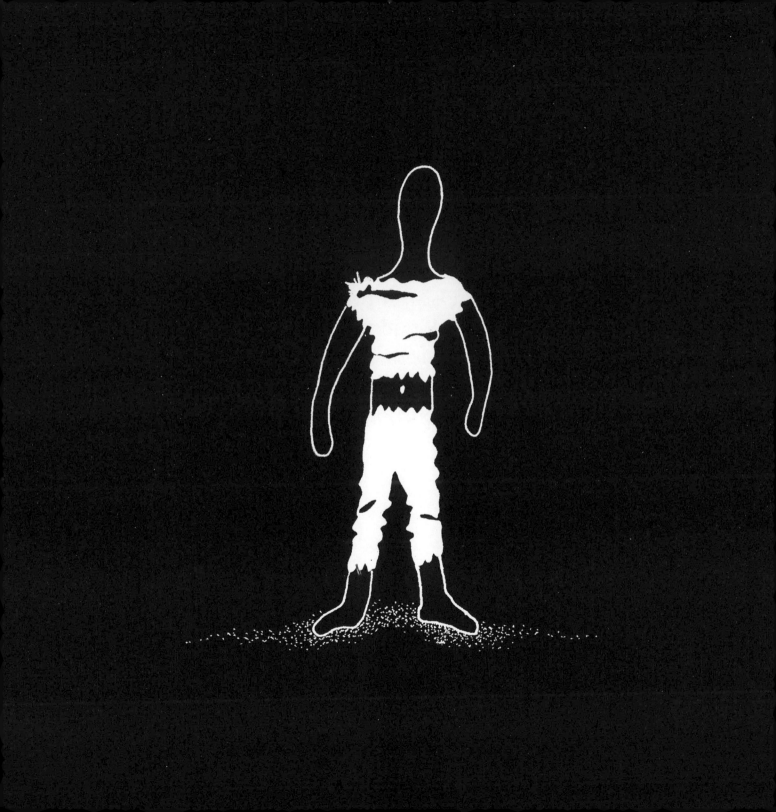

Why can't we see this? Why does our present agreement hide the truth? It hides it for the same reason that we can't go back and see the world through our ancestors' eyes. Because all agreements, all figurations, all forms, by definition, have limits. Throughout history, all agreements changed and expanded as the old agreements (paradigms) became too limited. Our present agreement is too limited! We have outgrown it, as is obvious by the discoveries of the quantum world, to say nothing of the dilemmas of our daily lives.

Like our ancestors who experienced the
beginning of the scientific revolution, we are living
at the time of a very big paradigm change. However,
this time, the new paradigm will incorporate within it all
of the previous paradigms. Why? Because our consciousness,
our figurations, all our paradigms are one, ever continuous,
expanding whole. And now, for the first time, we have the
ability, even scientifically, to know that. Consciousness IS
evolving and becoming more AWARE of itself. We are
this AWARENESS, and now have the ability
to see this evolution.

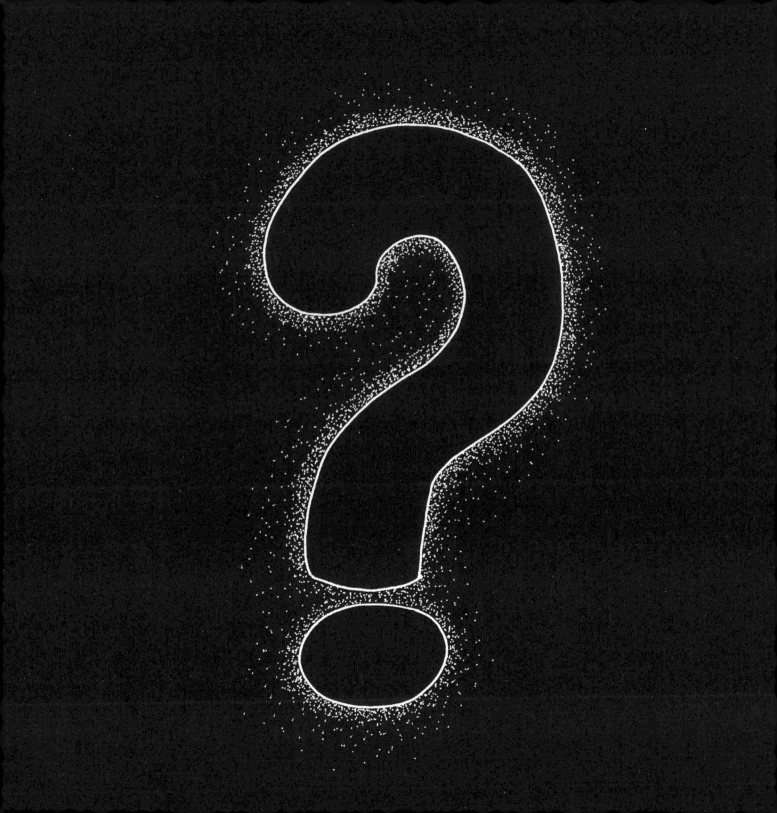

So LET'S SPECULATE! As you and I and all of us who make up this collective agreement, begin to recognize the huge significance of what we have been discussing here, what will happen?

At the very foundation of our present collective view is our perception of history. This will radically change. All collective agreements, including our present one, depend on our figuration of waves/particles into objects. That is because of the intimate relationship between those particles and our senses and our minds. If that is so, can we really view the objects in our historical past accurately using our present collective agreement? Are those objects static and unchanging? In other words, can the phenomenal world ever really be studied historically, or even objectively? Obviously, the answer is NO!

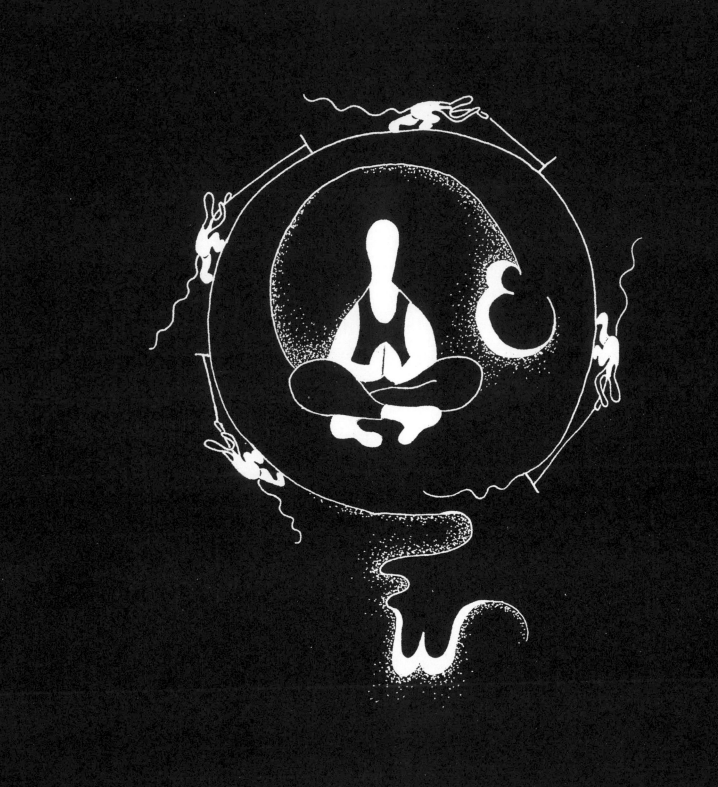

Does this mean that we can't study history?
No, of course not. But we can't study the history of objects, because all that we ever really study is our present (about to change) collective agreement. What we can study historically is the evolution of our collective views, and how that evolution is allowing us to become more conscious of what our beings really are. Our consciousness is Infinite Potential, Intelligence, Unfragmented Wholeness and Awareness, ever becoming more aware of itself.

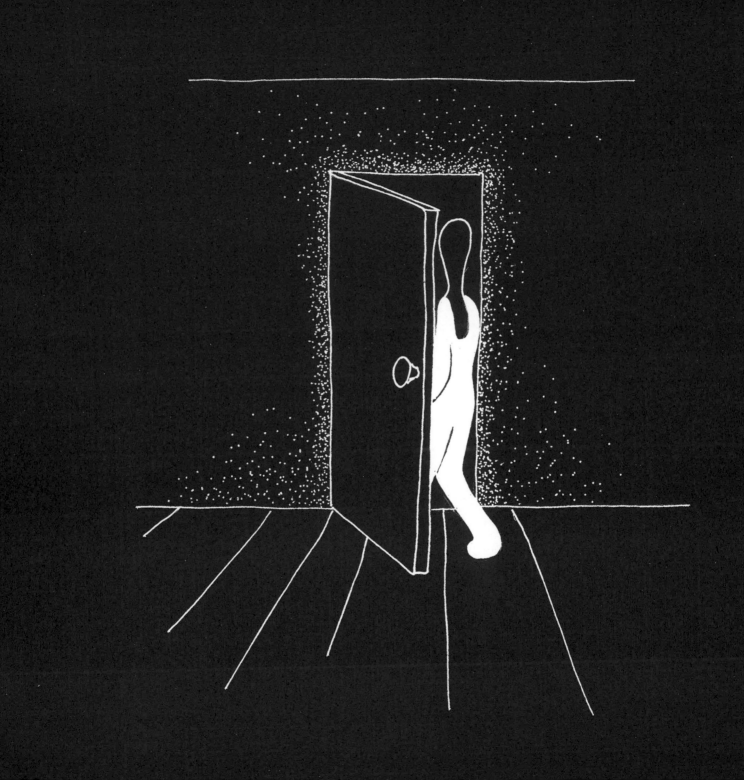

Can we end here, and just leave it at this?
We cannot, since we may not survive, historically or
otherwise, if we don't change our present approach—
to our planet and to ourselves. Many of us are actively
engaged in raising human understanding to a level that
will divert the disasters of global warming and other
catastrophes that would leave our earth uninhabitable.
Our huge misunderstanding about who we are affects
every single area of life on this planet. Let's take a further
look at who we are, where we are and where, perhaps,
we should be by now.

Here in Black and White

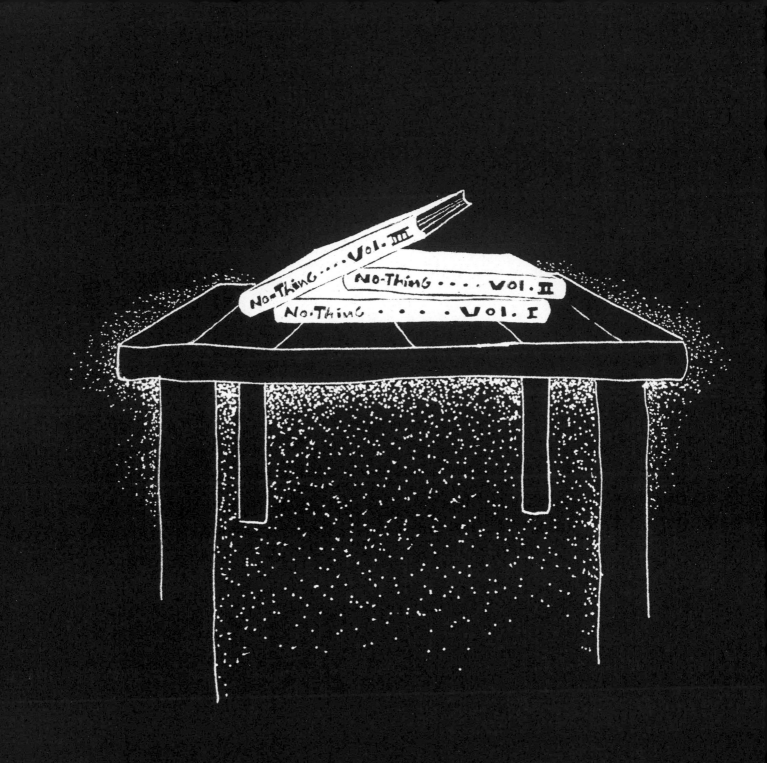

So here we have it: the last few chapters, all in black and white.

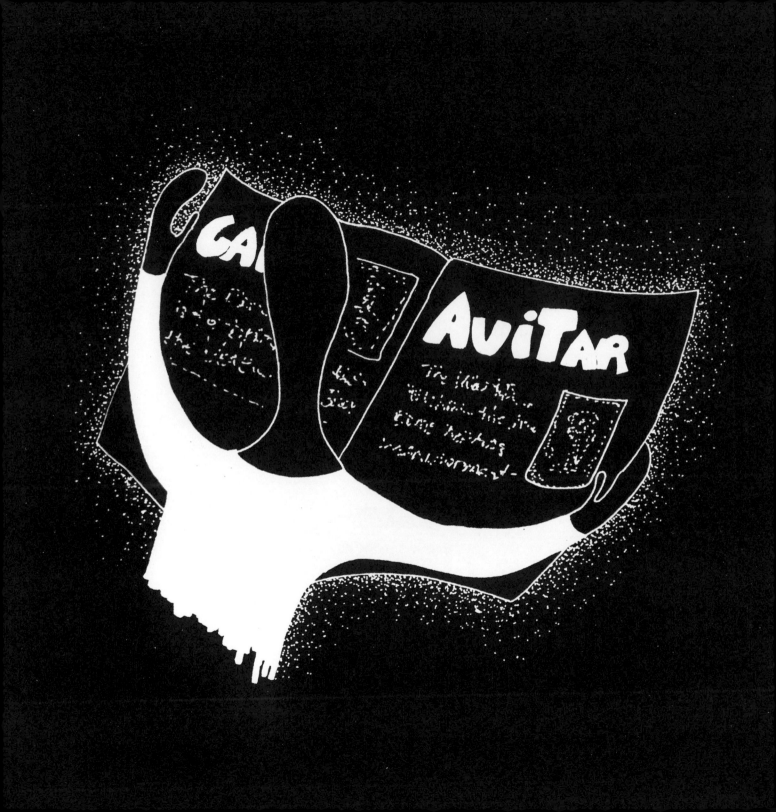

The very fact that we can read
those words depends on contrast of some kind.
We've all heard someone say, or even said ourselves,
"there it was in black and white."

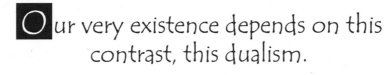

Our very existence depends on this contrast, this dualism.

Dualism was covered to a great extent at the beginning of the book. Then I wrote about how our senses and our minds shape our perceived reality.

Let us go a little further to better understand how critical this information is to our humanity and to our evolution as a species. It may even affect our continued existence as Homo sapiens on this planet. And, interestingly, it also affects our comfort level, our happiness and our peace.

First, recall that the measurable degrees or attributes of hot and cold, and the concept of temperature, are inseparable. If we have hot and cold—measurable attributes or degrees, then we also have the concept of temperature. If we have the concept of temperature, then we must also have measurable attributes.

LL

the countless pairs of opposites or contrasting elements that make up our lives are in this way the same. All combine measurable attributes with concepts.

P LUS

in every case, the concept that accompanies
dualistic pairs, organizes and arranges their 'parts' into a
useful and understandable whole. In the case of temperature
the parts are degrees. From now on, to emphasize this
characteristic, think of it as an 'organizing concept.'

INTELLIGENCE + PRINCIPLE

This organizing concept, by very
definition, implies intelligence of some kind,
and because it applies to ALL dualistic phenomena,
it qualifies as an observable Principle.

Turning again to the example of
temperature covered earlier, let's
examine for a moment the very important,
but often misunderstood and misused,
role of perception.

Have you ever found yourself waiting
for a bus or standing in line for a movie, arguing
with a friend about the temperature?

"It's freezing," you say, hugging yourself
to keep warm. "No, it isn't," replies your friend. "It's
very pleasant. Here, take my coat. I'm getting hot."
"But you will be cold," you argue. "No, I won't,"
says your friend, draping her jacket
over your shoulders.

It is not difficult to see that this argument is simply a matter of perception. Maybe you are just getting used to living in Phoenix in the summer. Or maybe your metabolism rate is slower than your friend's.

But you and your friend COULD get caught in a senseless argument, mixing up fact and perception. It happens all the time.

"It is 70 degrees today. I heard it
on the news. The fact is that 70 degrees
is warm," your friend states. "I don't care
about your facts," you respond, "that breeze,
coming from somewhere, is freezing.""Hey,"
your friend says, "I don't feel a breeze.
And 70 degrees is 70 degrees."

As human beings we often assume that our personal perceptions are fact. Our political and religious perceptions are obvious cases, and there are subtle ones as well—our personal views of morality or truth, for example. While 70 degrees may be a reportable fact, whether we perceive it as hot or cold is based on our perception.

That said; let's move back to the discussion of temperature as organizing concept and degrees as measurable attributes.

We have referred to the difference between Eastern and Western philosophies.

The Eastern philosophies often speak of form and emptiness. They tell us that the 'ground of all being' is both form and emptiness, and that they are the same thing.

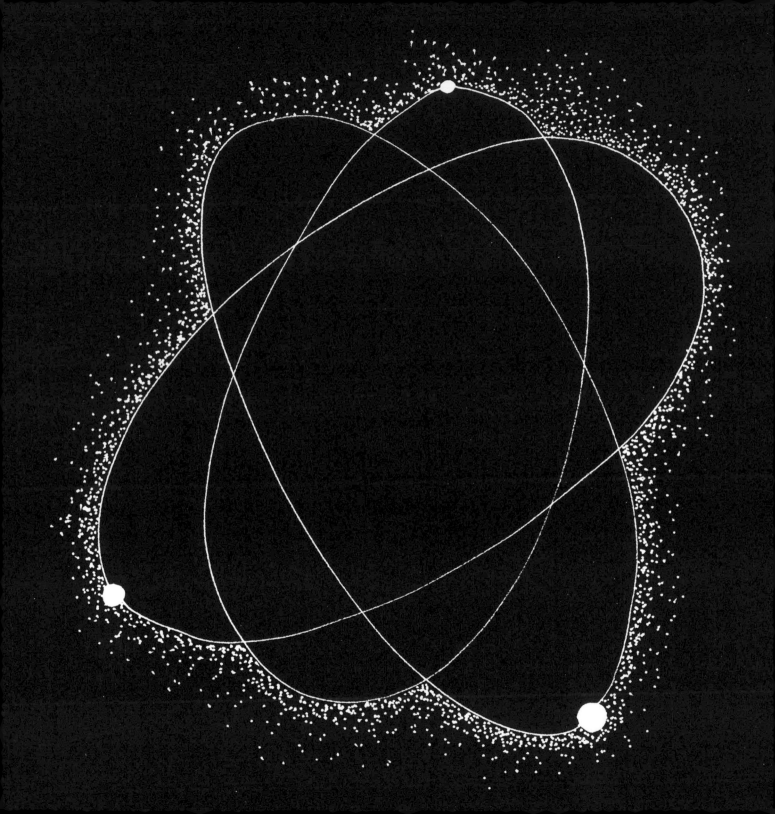

Our relatively recent Western science tells us that objects are not solid, but are composed mostly of vast empty space.

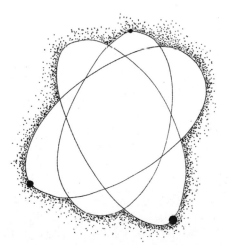

In simplified terms, quantum physics says that waves, which are potential, remain potential until observed, at which time they become particles, which are actual.

It seems that Eastern philosophy and Western science have met.

P revious chapters went into great detail concerning the 'mechanics' behind this interesting reality. I emphasized not only how pairs of opposites act as a conduit that enables consciousness to be, but also how the limiting aspect of our dualistic sense perceptions acts as a separating agent. Our dualistic sense perceptions allow for infinite individual experiences, as well as for the many different collective paradigms that fill our history.

And even though we may understand all
this on an intellectual level, it won't do us much
good until our understanding is experiential—
and can take its rightful place as
our PRESENT paradigm.

T hat is the whole point. The entire reason for this book is to help us all understand this looming shift, and to open the way for us to actually experience it.

Most interesting is that although we haven't experienced it, the shift has already occurred. Much of the technology that we use mindlessly every day would not exist without the discoveries of quantum physics.

If we stopped to take a look, we would see
that much of our daily discomfort, the time crunch
we all experience, for example, is due to the fact that
we have one foot in a belief system we do understand,
and one foot in a system we don't understand. Without
knowing it, we straddle the beliefs and attitudes of Newtonian
physics and quantum physics. We mix them, thinking of
ourselves in Newtonian terms as solid and separate, while
operating in the quantum world of computers, cell phones
and virtual reality. Blending these two realities has changed
our daily lives as much as or more than anything has
ever changed human lives in recorded history.

And to make matters worse, a good portion of the world that makes up our science and should be leading us toward a whole new understanding, does not itself understand the significance of its own discoveries.

WHY?

Because, many years ago, as we approached the era of the scientific revolution—resulting in Newtonian world accomplishments, we absolutely stopped asking the question, "Who are we?" The great minds of that era, assumed that that question was unanswerable scientifically, and thought it would get in the way of objective scientific research. They thought such a question should be left to the more subjective, philosophical or religious thinkers. Our quest became, instead, to expand our own power and to 'make nature do our bidding.'

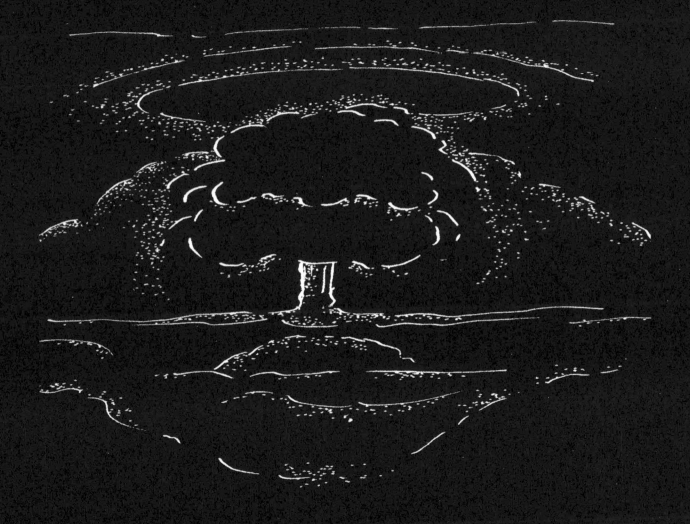

If you don't ask the right question, you won't get the right answer. Now the times have changed and the modern scientific community's failure to ask the question, "Who are we?" has lethal potential.

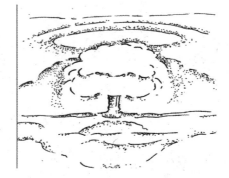

This out-of-date, blatant omission of the vital question that even the youngest of us has sense enough to ask, has placed us on the brink of our own destruction.

et us all ask it now.

WHO ARE WE?

Who are we? And how will the analogy of temperature, picked out of the infinite number of pairs of opposites, help us?

How will that duo of 'concept and measurable attributes' that we deemed an 'observable Principle' help us know who we are?

It helps us because that 'duo' is a 'Principle' and that Principle applies to all pairs of opposites, all contrasting elements, everything dualistic.

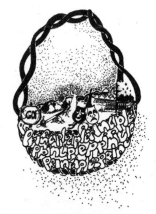

UT

this duo is the Big One:

IT APPLIES TO FORM AND EMPTINESS

The East has always talked about form
and emptiness. The West now finds it an intricate
part of its science (waves and particles). Both systems
of thought would have to concur that form and
emptiness are the 'ground of all being.'

Our use of temperature is an example of the observable Principle. Is not temperature also an example of form and emptiness? The organizational concept of temperature is not measurable, is not a thing, and is therefore empty. Degrees of temperature are measurable, and therefore have form.

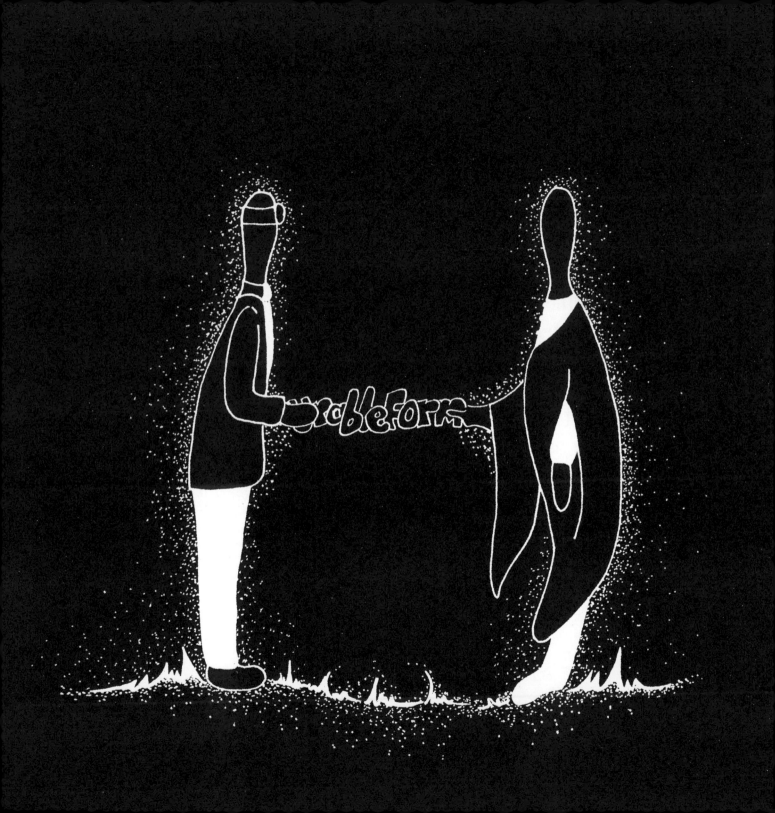

MEASURABLE AND IMMEASURABLE ARE ALSO 'FORM AND EMPTINESS'

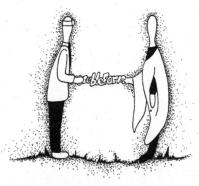

Besides all this, quantum physics has shown us that FORM is not so solid, and EMPTINESS is not so empty.

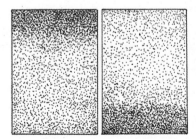

A nd since our world is dualistic,
we live and move around in an environment
made up entirely of

EMPTINESS AND FORM

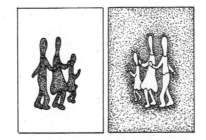

To reiterate, all the contrasting elements of this world that are measurable are objects, which also have immeasurable organizing concepts. The measurable and the immeasurable are inseparable. The Principle that is the 'ground of all being' is that 'form and emptiness— measurable and immeasurable' are two aspects of the same thing.

That Principle applies to us too. Each of
us is, individually, both a measurable body living
in time and space, and also an immeasurable organizing
concept. This organizing concept is clearly known to us.
IT IS INTELLIGENCE!
The very definition of an organizing concept
implies intelligence of some kind.

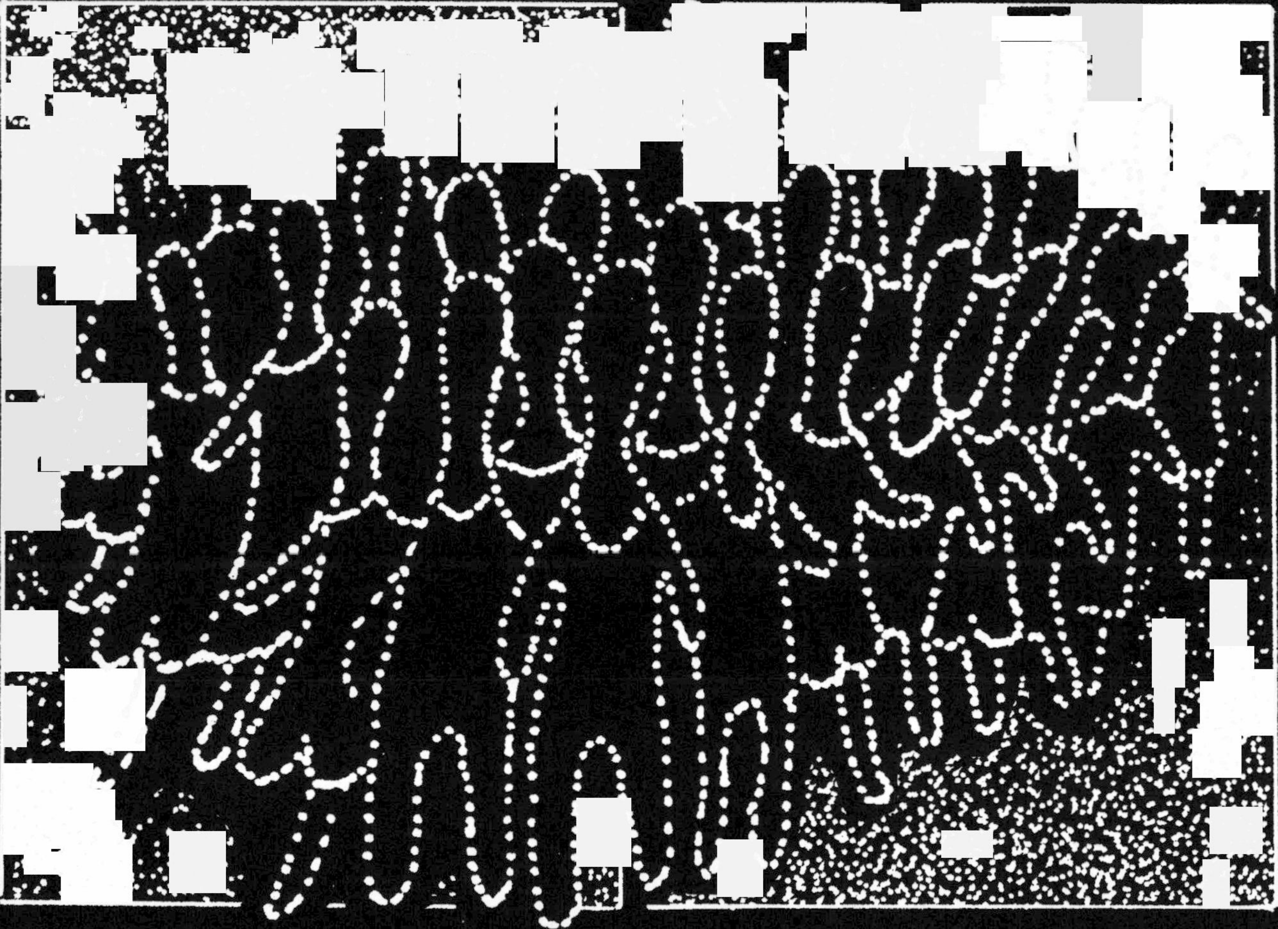

That Principle also applies to us collectively as a human race, as well as to everything known to us.

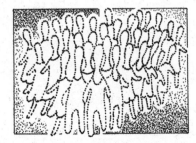

Over and over in these pages, I have
shown examples that even measurable objects
are not made of material stuff. I have shown that
measurable and immeasurable are actually
two different aspects of the same thing.

Whether it is a seed we planted, growing in our garden, or a rock we use as a paperweight, or even a plastic water bottle we take on a walk, all these objects are products of and dependent on an organizing concept. That organizing concept has as its basis intelligence, and is this Principle.

Let us now ask some very pointed questions.

In order for this intelligence to manifest into things, must it not have a focus of attention? Do not our sense perceptions and our minds, as described in earlier chapters, provide that focus? Is this not the same focus of attention that the scientist uses to observe that wave that becomes a particle? And are his instruments not an extension of his senses?

When this scientist and his instruments focus
on that wave, is the scientist himself not an example
of our collective human potential? Does the very fact
that he has a body with a unique set of sense perceptions,
not make it impossible for him to be an independent observer?
Like the argument debating whether 70 degrees is hot or cold,
is the scientist not limited by the 'tool' that is his body?
Furthermore, is he not using a mind that is part of a
collective paradigm, based on his fraternity of fellow
scientists, his religious affiliation, on and on—until
his personal views combine with the
present collective paradigm?

Are we caught in a squirrel cage? Are we running around and around in a system of our own making, only to break out of it every few centuries to create a new, but just as confining, cage to run around in—AGAIN?

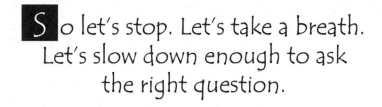

So let's stop. Let's take a breath.
Let's slow down enough to ask
the right question.

WHO ARE WE?

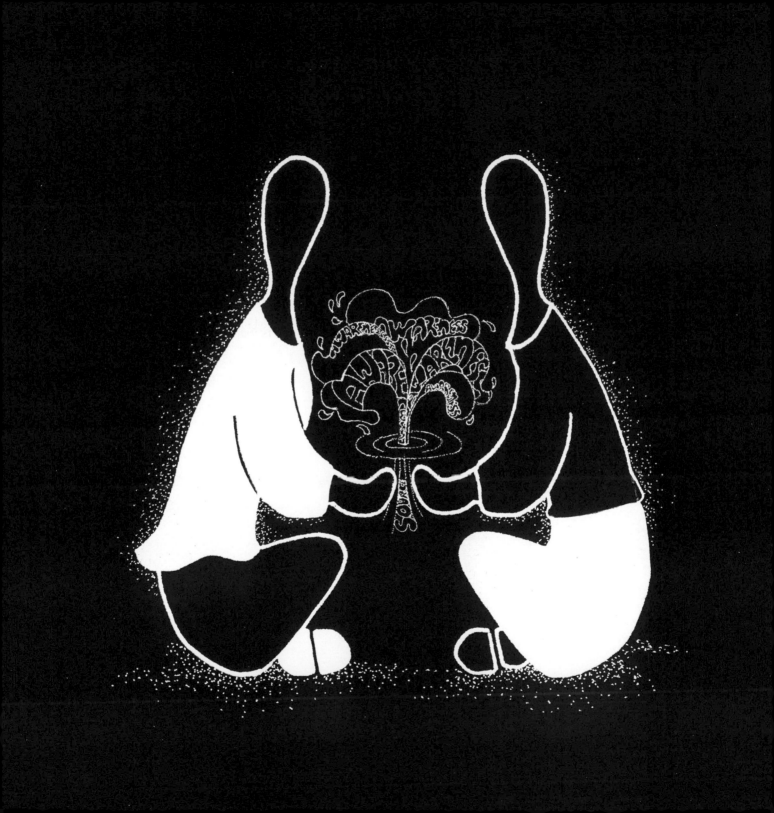

Coupled with the wealth of current and timeless information, can we be very very still? Can we silently ask the question, "Who are we?"

With all that we know now, can we listen for that intelligence, that organizing Principle, that we know that we are?

Can we be still enough to realize that when ALL else is silent, what is left is the 'ground of our own being,' and of all beings from which ALL becomes manifest?

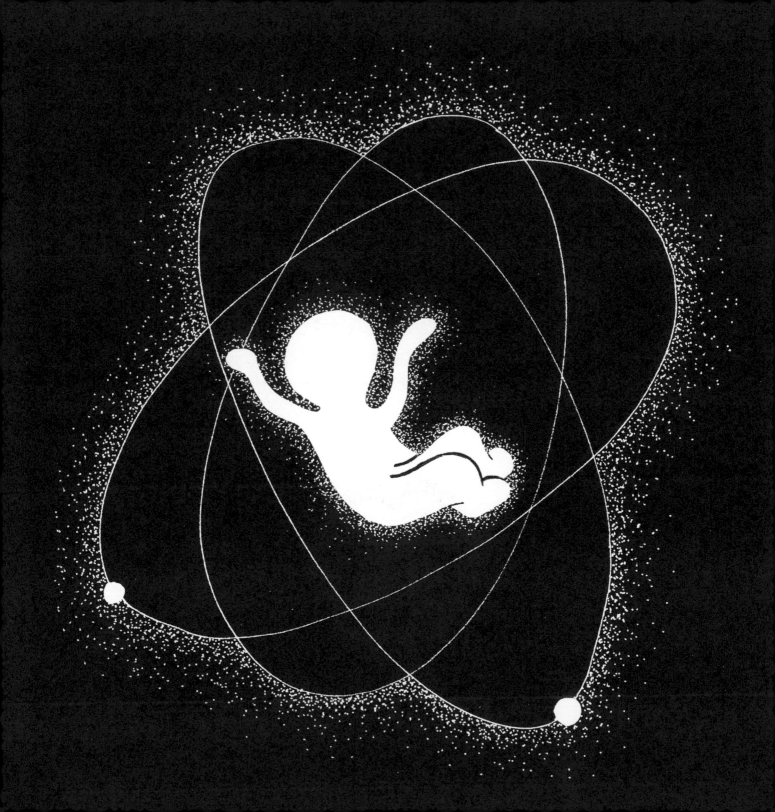

And from the silent depths of our beings, can
we feel what the next action should be? Can we tap
into our own supremely intelligent, organizing Principle—
the same Principle of intelligence smart enough to grow a
baby, smart enough to create the entire solar system? Are
we mature enough to get over our own importance and
step into the truth of our beings—that are not separate
from, but are an integral part of the whole system?
Are we ready to stop suffering?

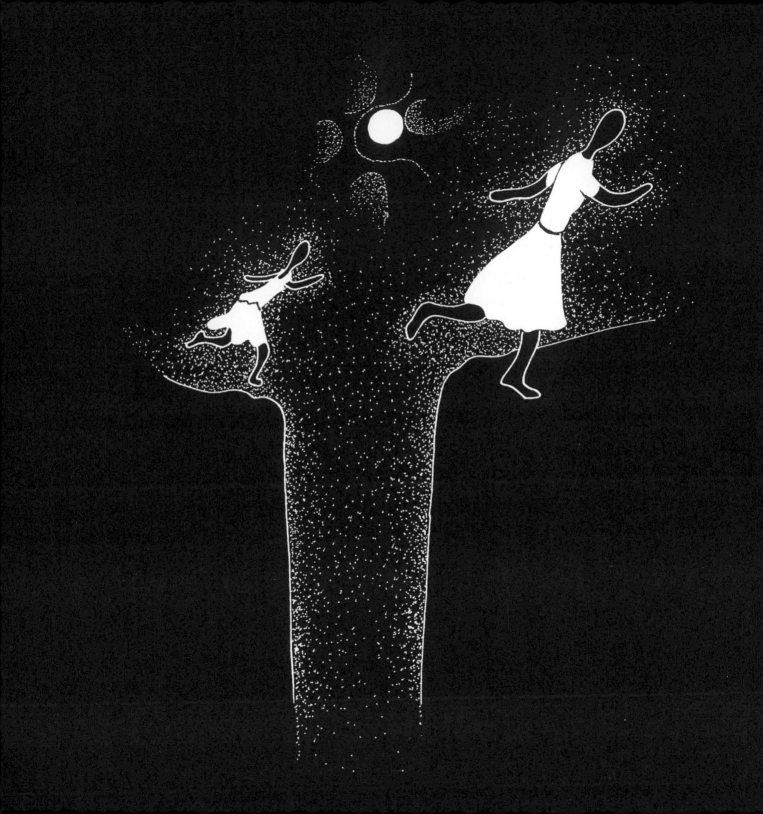

In truth, our time is up. Our childhood is over. If we can't make this simple leap that requires absolutely no faith at all—due to our quantum discoveries, then we are truly expendable and unworthy.
HOWEVER

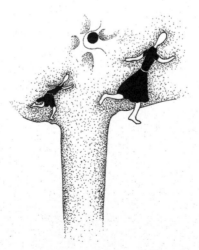

One way or another, thanks to the organizing Principle itself, this ever-expanding evolving consciousness that we are will continue expanding … growing …

For …
EXPANSION IS ITS VERY NATURE!

ACKNOWLEDGEMENTS

As I wrote and illustrated this unconventional book, I had no idea that I would publish it. I had no idea how to publish it. When my good friend, Alexandra Shamaya, an author-publisher herself encouraged me to consider this possibility, it was her energy that accompanied this invitation that sealed the deal. Without Alexandra this could not have happened. She put me together with James Bennett, a book designer of extraordinary talent, who infused professionalism into my work. She introduced me to Greg Brandenburg, Associate Publisher at Hampton Roads Publishing, who from my point of view, miraculously brought this effort across the finish line. A serendipitous coincidence involving Debbie Harris Funfer, my dearest friend and indispensable-beloved critic, brought forth Karen Meadows, my editor. Karen unobtrusively sat with me and helped shape my words in such a seamless way that I can hardly detect the changes. To you all I wish to acknowledge my deep indebtedness.

To my family I extend my undying gratitude. Without you, Leaf and Lorenz, and your curious minds and life questioning attitudes, there would have been no inspiration to even begin this project. Ramona and Brian, your unwavering support counted more than you will ever know. Charlie, you contributed to this endeavor in ways too countless to even name. And last but not least, it is you Bailey, who enthusiastically push all of us forward to meet a future that is brimming over with potential.

Thank you all from the bottom of my heart.

Author and illustrator,
Terry Favour

Terry with her granddaughter

When the author, Terry Favour, first asked the question "Who Are We?" her very ability to form such thoughts and sentences was just beginning. The failure of those around her to answer back, set her on a quest that would last indefinitely. The process led her across a diversified terrain, weaving back and forth through spiritual philosophy and science, from East to West.

As an adult she has sustained herself and this passion working as an artist and designer, owning and operating a small ethnic arts company with her son, located in Northern New Mexico. In the 1980s she became interested in the Enneagram. In 2004 she and her daughter studied under David Daniels and Helen Palmer and they both became certified Enneagram teachers.

She feels that expanding one's limitations should be an unending endeavor so she continues this life work every day. It was her grown children who led her to write this book, stating that she had a knack for simplifying complicated matters.

She continues to draw and write and she lives in Santa Fe, New Mexico with her husband and her cat.